Petroleum Engineering for Kids

Ahmed ElGamal

Yassin: Hey daddy, tell me about your work as a PETROLEUM ENGINEER.

Dad: Oh my little Yassin, you are growing up so fast!
Close your eyes, let me tell you all about the oil industry.
Yassin

Yassin: What is oil used for?

Oil makes petroleum products which help us in our everyday lives. We use them to power our cars, fuel our trains & planes and heat up our houses.

Oil byproducts are also used in the production of plastic which is what all your toys are made of!

Yassin: Where did Oil come from?

Millions of years ago, there were dinosaurs and plants that lived on Earth. When those animals and plants died, they got buried and were covered with new layers of earth, layer upon layer. The same happened with algae & plants living in the sea.

When they died, they sank to the seafloor where the remains got buried deep into the earth's crust.

Then, over millions of years, those buried plants and animals were heated to extremely high temperatures under lots of pressure from the layers on top of them. They got 'cooked' into thick, black hydrocarbons.

Yassin: How do you find the oil?

Hydrocarbons never stay still. Over time, they move through pore spaces in the rocks. Hydrocarbons get trapped by rocks that have tiny pore spaces. When the rocks trap the hydrocarbons, they seal them in reservoirs. It is nice to think of hydrocarbons' reservoirs as a huge sponge with oil bubbling through its holes.

Geoscientists are clever people who study the Earth's underground by producing special drawings called seismic maps. Those drawings enable the geoscientists to look for those traps & spongy reservoirs which store the hydrocarbons. When they find them, the petroleum engineers poke the sponge with a big, tall straw from the surface to get that oil out.

Yassin: How do you get that oil out?

As I said, the petroleum engineers
try to poke the sponge that holds the
hydrocarbons with giant straws from the
Earth's surface.

To get these straws into the ground, the petroleum engineers use a drilling rig with straw-like pipes to reach the reservoir. Once a hole to the surface is created, the hydrocarbons can move upwards due to the difference in pressures from the top to the bottom.

How do you transport oil?

Oil is transported through big trucks, sea vessels, on railroads and through pipelines.

The process of transporting oil can, a lot of times, cause some environmental catastrophes. That is why we must make sure that the oil is safely transported from one place to another with no harm to humans, animals or our lovely environment.

Yassin: What happens to the oil then?

After hydrocarbons are transported to those big refinery sites, it is then time to get each hydrocarbon product separated and cleaned, which is called oil refining. Heavy parts of petroleum separate on the bottom, while light petroleum products move towards the top. Some of the many products that result from this process are jet fuels for fueling our airplanes, and asphaltene for paving materials on roads.

Yassin: What else should I know about oil?

Well, the whole process of exploring, drilling and producing oil can hugely harm our environment. Burning hydrocarbons produces CARBON DIOXIDE, known as greenhouse gas, which is the main cause of global warming.

Sulphur Oxide is also another negative byproduct of burning hydrocarbons which causes 'acid rain' and is extremely harmful for humans, animals and plants.

The production of oil is crucial in making our lives easier as humans. As I have told you, it is pretty much involved in the production of almost everything. However, we must come up with a way of producing oil while keeping our environment safe from the contamination of the byproducts.

Yassir

Yassin (hugging his mom): "Hey mom, now I know everything about oil and when I become a petroleum engineer, I will make sure to take care of my environment too!"

What is oil used for? _______________________________

____

____

____

Where did oil come from? ___________________________

____

____

____

How do you find oil? _______________________________

____

____

____

How do you get oil out? ____________________________

____

____

____

How do you transport oil? _______________________________

____

____

____

____

How is the oil refined? _________________________________

____

____

____

How is oil related to the global warming? _________________

____

____

____

I must first acknowledge my wife for her patience & understanding; oh yes, I spent many hours away jumping In front of my computer screen as I got every page ready. I would also like to thank my parents who threw me on the rig while I was still young, as they cant stand my sense of humor at home. Finally, I would like to remind my little son that I love him even if he doesn't become a Petroleum Engineer.

The author, Ahmed ElGamal, born in 1991, has been involved in the Petroleum industry for more than 13 years. Jumping from one rig to the other, AG has built enough experience to at least describe what a rig looks like to a 8-year-old child. Parents and children please get your safety gears on as we jump into the world of PETROLEUM ENGINEERING FOR KIDS.

Disclaimer: Don't blame the author if your kid turns into a future petroleum engineer